Ho

the WRONG PLACE at the WRONG TIME

Robert Gott

Published by Sundance Publishing
33 Boston Post Road West
Suite 440
Marlborough, MA 01752
800-343-8204
www.sundancepub.com

Photography
Cover Jupiterimages Corporation

First published as Phenomena by Horwitz Martin
A Division of Horwitz Publications Pty Ltd
55 Chandos St., St. Leonards NSW 2065 Australia
Exclusive United States Distribution: Sundance Publishing

ISBN 978-1-4207-0731-1

Printed by Nordica International Ltd
Manufactured in Guangzhou, China
April, 2014
Nordica Job #: CA21400466
Sundance/Newbridge PO #: 227714

Contents

Preface

Have you ever had one of those days when strange things happen to you? Maybe you were thinking about a song when suddenly that song is playing on the radio. Maybe you had a test that day at school. You hoped and hoped that your teacher would be out. Then when you arrived you found out that your teacher really was out. That's not really a coincidence, but it is good luck (not for the teacher, of course).

Everyone has had some experience of amazing coincidences. Some of them are small, but some of them are really astonishing. The coincidences you will read about in this book are not your everyday, run-of-the-mill coincidences. You won't believe that they could possibly have happened.

You might have trouble believing the stories of survival, too. They are, however, true, and they happened to ordinary people, just like you. These stories show us how extraordinary we can be when we are tested to the limit.

This is a book about chance. It's about the tricks that chance can play on us. Chance puts some people in the wrong place at the wrong time. Other people are in the right place at the right time. Individuals deal with the cards they have been dealt in different ways. This book tells the stories of some of those individuals.

1 Introduction

Imagine: going to meet a twin brother you never knew existed.

Now I know, before I even begin this story, that you are not going to believe it. I barely believe it myself, and it happened to me. This isn't one of those little coincidences. Not like when you're about to call a friend and he calls you first. Oh no. This is quite possibly a record-breaking coincidence. It's so bizarre that you'll think I'm lying. I'm not. In fact, I can prove that I'm not.

If you look up the newspaper for the day after this happened, you'll see it reported there. Of course, you'll have to get the *New York Times* because that's the newspaper I'm talking about. Or you can look it up on the Internet. If you can't, you'll just have to take my word for it.

I'm 14 years old. Big deal. I'm an only child. Also, big deal. Well, it is a big deal. On the day I turned 14, my parents told me that I had an identical twin brother. I don't think you can imagine how that freaked me out. All of my life I had gone along thinking that there was only one of me. Now suddenly I hear that there's another me.

The details of how we came to be separated are important, but they're private. To put it bluntly, they're none of your business. Mom and Dad explained it all to me, and I understood the position they were in. That's all I'm saying.

Anyway, when I turned 14, they thought it was time I knew the truth. They'd made this deal, you see, to tell me then. Somewhere on the other side of the country, my twin was being told exactly the same thing. My parents had kept in contact with the people who had adopted Harry—that's my brother's name. They confessed that they had been dreading this day. They didn't know how I was going to react. They especially didn't know how I was going to react to the special news. My birthday present was a plane ticket to New York City to meet my brother. I pretended that I was excited. Actually, I was terrified.

We were supposed to be met at the airport by Harry and his parents. After we waited for an hour, Dad decided that they must have gotten the date wrong. He decided that we should catch a taxi and surprise them. I thought it would be more sensible to call and find out what had gone wrong. I've always been more sensible than my father. He insisted that we take a taxi. Here's where it gets really weird.

We were turning onto the highway when a car came out of nowhere. It slammed into the side of our taxi. Luckily, we were wearing seat belts, so we weren't thrown around the car. The noise was terrible. Is there a worse sound than grinding metal and breaking glass? Then there was this awful silence.

"Is everyone okay?" Dad asked.

No one was hurt, just shaken up. The taxi driver was angry because the other driver was at fault. Or at least that is what he thought. He started yelling at the driver of the other car. We all got out of the car, and the people in the other car got out, too. We were all a bit dazed.

There were three people in the other car. I couldn't see them clearly because they were on the other side of their car. I noticed that one of them was a boy about my age. When he came around from behind the car, my heart actually stopped beating for a moment. There in front of me was a mirror image of myself. I had come all the way to New York City to meet my twin. And we had crashed into each other.

I knew you wouldn't believe me. You should. Just ask Harry. . . .

1 Incredible Coincidences

When was the last time that you stopped and thought, wow, that's a strange coincidence?

Our lives are full of coincidences. Some of them are small, and some of them are really weird, like the coincidence in the story. There are many, many strange stories involving identical twins. If you collected all of the coincidences that have happened to you or your friends, you would probably have a book. Most people love coincidence stories. But how people make sense of these coincidences can be quite different from one another.

A famous psychologist named Carl Jung studied coincidences. He believed that coincidences revealed that in some way we were all linked to a universal force. He called his theory synchronicity.

He believed that all coincidences had meaning, that they were more than just random accidents.

Not everyone agrees with Carl Jung. Many people believe that coincidences are simply examples of chance. Whatever you think about coincidences, there is no doubt that they fill us with wonder. Sometimes they even change the way we see the world.

The One About the Missing Jewelry

The world is full of stories about small, valuable objects that have been lost. And then they turn up in the oddest places. It's one of the most common tales of amazing coincidence. You've probably heard a story similar to this one. A fisherman's wife lost her expensive earring while leaning over the side of her husband's boat. A few days later, she was cutting up a large fish caught by her husband. And there was her earring, shining in the fish's belly.

How about this one? A man lost his wedding ring during a flood. It was swept away. He could not replace it, and he was very upset. A few days later, a local chef was cleaning a fish and found the wedding ring.

There are many, many stories of coincidences like this. Many of them are a bit questionable. Is it possible that they are not all they seem? We can actually trace the jewelry in the fish story all the way back to ancient Egypt. In 550 B.C., a tyrant named Polycrates was terrorizing the people who lived in his area. The pharaoh decided to do something about it. He dared Polycrates to test his fortune by throwing a valuable ring into the sea. If fortune was on his side, the sea would return his ring. A few days later, a fisherman brought Polycrates a fish as a gift. The ring was found inside. This fish story is remarkably similar to modern stories.

Edgar Allan Poe

Edgar Allan Poe wrote what is considered to be the world's first detective story. It was called *Murder in the Rue Morgue.* Many of his stories are spooky. In 1838, he published a novel called *The Narrative of Arthur Gordon Pym.* In Poe's story, there is a shipwreck, and the survivors face starvation. In typical Poe fashion, their solution is horrible and meant to shock the reader. The survivors decide that they will kill a cabin boy named Richard Parker and eat him. The story was very popular. Incredibly, 46 years later, an odd event took place that reproduced this story almost exactly. This was long after Poe had died, and his book was no longer widely read.

In 1884, a ship called the *Mignonette* went down at sea. Many people were killed, but there were a few survivors who managed to get into a small boat. They floated around aimlessly for several days. Thirst and starvation made them desperate. There were three men and a cabin boy in this boat. The men decided that they would kill the boy and eat him. The boy's name? Richard Parker.

Maybe He Should Just Stay Home

Have there ever been times in your life when you wished that you had just stayed home? It seems that some days nothing goes right. When you have one of these days, think of a man named Joseph Figlock. One day he was walking down a street in Detroit, minding his own business. Suddenly a baby, who had fallen from the 14th floor of a building, fell on top of him. He broke the baby's fall, and no one was badly hurt. That's one lucky baby. A few seconds either way and it would have hit the pavement. Imagine the story he told his family that night. One year later, Joseph Figlock was walking in Detroit again. As he passed a building, another baby fell on him. No one was badly injured. It makes you think, doesn't it?

James Dean's Car—Is It Haunted?

James Dean only made three films, yet he remains one of America's most famous movie stars. At the age of 24, he died in a terrible car accident. He was at the wheel of his car when he crashed into a tree at high speed. He quickly became a legend. Anything associated with him became valuable, including the wreck of his car.

James Dean's car was too valuable to be sold for scrap, so it was repaired. Several strange coincidences followed. So some people believe the car was haunted, or jinxed. Soon after the crash, a mechanic was working on the car. He suffered two broken legs when the engine fell on him. Such an accident is not too unusual among mechanics. Two doctors, however, bought parts of Dean's car to use in their race cars. On one occasion, one doctor was killed in his car. The other doctor was seriously injured, also when driving his race car.

By now you'd think that people might want to stay away from parts of the car. But there was, and still is, a lot of interest in James Dean. What was left of his damaged car was put on display in Sacramento, California. It attracted big crowds. One day it fell off its stand for no apparent reason and broke someone's hip. Soon after this, the car was moved to another location on the back of a truck. The truck crashed through a store window. In 1959, the car finally destroyed itself. It was on display again, and like it did before, the car fell off its stand. This time it broke into 11 pieces.

If I ever bought a car from the 1950s, I think I'd check one thing. I'd make sure that none of its parts came from James Dean's car. I'm not really superstitious, but some things are too weird to take a risk.

John F. Kennedy and Abraham Lincoln

When people think about coincidences, they often look for connections that are mysterious. It's as if they want to believe that coincidence is more than just chance. They want to believe that weird forces are making things happen. Many coincidence hunters have looked at the similarities between President John F. Kennedy and President Abraham Lincoln. They have found a number of amazing coincidences between these two presidents. Remember, these events took place a century apart.

General Coincidences

- Abraham Lincoln was elected to Congress in 1846. John F. Kennedy was elected to Congress in 1946.
- Abraham Lincoln was elected president in 1860. John F. Kennedy was elected president in 1960.
- The names Lincoln and Kennedy each contain seven letters.
- A child of Mrs. Lincoln's died during her stay at the White House. A child of Mrs. Kennedy's died during her time at the White House, also.

The Assassinations

- Both presidents were shot on a Friday.
- Both were shot in the presence of their wives.
- Both were shot in the head.
- Both presidents were killed by Southerners.
- Both presidents were succeeded by Southerners, and both men who took over the presidency were named Johnson.
- Andrew Johnson, who succeeded Lincoln, was born in 1808. Lyndon Johnson, who succeeded Kennedy, was born in 1908.
- The man who assassinated Lincoln is known by his three names, John Wilkes Booth. The man who assassinated Kennedy is known by his three names, Lee Harvey Oswald. Each name contains 15 letters.
- In an attempt to escape capture, Booth ran from a theater and hid in a warehouse. Oswald ran from a warehouse and hid in a theater.
- Lincoln was shot in Ford's Theater. Kennedy was shot in a Ford Lincoln car.
- Booth was born in 1839, Oswald in 1939.

- Lincoln had a secretary named Kennedy, who warned him not to go to the theater that night. Kennedy had a secretary named Lincoln, who warned him not to ride in an open vehicle.

This does seem incredible, doesn't it? Can all these bizarre similarities be no more than an accident? Let's look a little more closely at some of these. It is not really that unusual for people to share the same birthday. The chances of it happening are pretty good. If you check among your classmates, you'll probably find two people born on the same day. The date of birth given for Booth in this list is wrong in any case. He was born in 1838, not 1839. Sometimes coincidence hunters stretch the truth a little where facts don't quite fit.

The coincidence about the secretaries is often repeated. This is because it seems truly amazing. The only problem is that it's not true. Lincoln did not have a secretary named Kennedy. He had two secretaries, John J. Nicolay and John M. Hay. No Kennedy.

Booth hid in a barn, not a warehouse. But to add interest to the story, it is almost always called a warehouse in the coincidence list.

There's no denying that many of the coincidences between the two presidents are amazing. However, amazing coincidences can be found almost anywhere if you look hard enough. In 1992, a contest was set up to look for coincidences between pairs of presidents. The winner found 16 coincidences between President John F. Kennedy and the former president of Mexico, Alvaro Obregon. Another person in the contest found coincidences among 21 different pairs of presidents!

2

Introduction

Imagine: Lightning can do more than light up the sky.

My grandfather hardly ever spoke to me. He was a tall man with sharp whiskers that scratched my face when he leaned down to kiss me. I loved him.

It wasn't just me that he didn't speak to. He didn't speak to anyone much. He sat on the porch outside his house, smoked his pipe, and watched the sky for passing birds. We visited him less and less as I got older. My parents got tired of the two-hour drive out of town. But when we did visit, I would sit at his feet. He would name each bird as it flew by.

"Crow," he would say. "Robin. Swallow."

I learned a lot about birds but not much about my grandfather.

The reason I'm telling you about him is that something amazing happened one day. I was sitting with him when it occurred.

"Have you ever been overseas?" I asked him.

"No," he replied.

"What was Grandma like?"

"Loyal."

Our conversations were like that. I would ask a question, and he would answer with one word. Dad grew up with this, so he didn't stick around much when we were at Grandad's. He went off with Mom down to the dam for a swim. Or they took a long walk in the woods. I liked Grandad's answers. They were reliable.

"Have you ever had a really bad accident?"

"Yes."

"What happened?"

"Car."

Did his eyes look sad when he said that? I thought so, but I knew better than to ask for more information.

He pulled his pipe from his top pocket and tapped it on his knee. He had a ritual that he always went through with his pipe. He tapped it, sucked on it loudly, and then began packing it with tobacco. He pushed the tobacco down with his thumb. Then he lit the match. The flame curved over and into the bowl of the pipe as he drew in his breath. Clouds of perfumed smoke would then come from his mouth.

On the day that I asked him about the accident, he began his pipe ritual as usual. The horizon was dark with clouds. They were a deep, threatening blue. A storm was on the way. There were occasional flashes of lightning. Not forks of lightning, just white highlights. There was no thunder.

I noticed that Grandad was watching the sky. He put the pipe to his lips, but he did not reach into his pocket for the matches. He waited a moment. That was when it happened. There was a flash of white light and a peculiar sound. I don't know how to describe it. It wasn't like thunder. There was a tearing sound right next to my ear. It was as if the air itself had been ripped like a cloth. I closed my eyes, just for a second, against the flash.

When I opened them, my grandfather looked completely amazed. His pipe was still in his mouth, but it was smoking. The amazement changed into a smile. It was as if he had made up his mind about something. He looked past me to the darkening sky and said, "Thank you."

I didn't bother telling my parents. The day my grandfather's pipe was lit by lightning became a secret we shared. This is the first time I have shared it with anyone. . . .

2

What Are the Odds of That Happening?

A bolt of lightning igniting a man's pipe? Surely not. In fact, the story of the boy and his grandfather's pipe is based on a real story.

The circumstances were different, but the man's pipe was struck by lightning, and he was unhurt. Have you ever had something really odd happen to you? Many people have experienced things that are so strange that they are almost unbelievable. All of them involve an element of chance.

Where Lightning Strikes

How often have you heard the old saying, lightning never strikes twice in the same place? It isn't true, of course. Tall buildings, such as the Empire State Building in New York, are struck many times each year. They are fitted with lightning rods that prevent the electric charge from doing any damage.

Lightning is a dangerous force. There are lots of myths about it. One of these is that you have to be outside in a storm to be in danger. In fact, half of all the lightning deaths in the United States occur indoors. It can happen when lightning hits the roof or the chimney. Then it strikes people who are near electrical appliances or fireplaces. This charge is called a side flash. It is also not a good idea to be on the telephone during an electrical storm.

Introducing Roy C. Sullivan

Not everyone who gets hit by lightning dies. More than 80 percent of victims recover. One man holds the record for being hit by lightning more times than anyone else. An ex-park ranger from Virginia named Roy C. Sullivan had a reputation as a human lightning rod. He was struck by lightning an incredible seven times. He survived each strike with only minor injuries.

Roy was struck by lightning the first time in 1942. He lost a toenail on that occasion but was otherwise unhurt. Little did he know that this was not the last shock he was to receive. It was quite a while before Roy was struck again. In July 1969, a bolt of lightning singed his eyebrows. He must have thought it was pretty weird to be hit twice by lightning in a single lifetime. It's a safe bet that none of his friends could make that claim.

The following year, in July 1970, Roy was struck again. This time, he was badly injured. Lightning delivers more than a little electric shock. It burns the skin. It's like having a blow torch turned on your naked flesh. Roy suffered burns to his left shoulder. By this time, he must have become nervous whenever rain was predicted. Then, on April 16, 1972, lightning set his hair on fire and blistered his scalp. How must it have felt to have been struck four times? He probably had trouble finding a golf partner. (It is actually not true that most lightning strikes occur on golf courses. Only about five percent of strikes occur there.)

Roy's hair grew back but not for long. On August 7, 1973, yet another bolt of lightning set his hair on fire. It also burned his legs. The following year, Roy escaped unstruck. He was safe in 1975, too. He must have thought that his run of bad luck was over. It wasn't. On June 5, 1976, lightning injured his ankle.

The Human Lightning Rod

What do you do if you feel like you're a lightning rod? Roy had to live his life as normally as he could. He took precautions, but there's only so much a person can do. On June 25, 1977, Roy decided to go fishing. This was a big mistake. He was struck by lightning for the seventh time on that day. This time he suffered burns to his chest and stomach. He must have been sick of the horrible smell of burning flesh and the terrible pain afterward. This was the last time he was hit.

There is no scientific explanation for Roy's extraordinary record. Perhaps he was just in the wrong place at the wrong time—seven times!

Oh, Lucky Man

Everyone hopes to get lucky when they buy a lottery ticket. But one man was luckier that most. At a time in his life when he really needed the money, he bought a scratch lottery ticket. To his astonishment, he won $100,000.

Because this was such a large amount, it attracted the attention of the local media. The winner agreed to reenact buying the ticket, and the moment when he discovered his win. He returned with a camera crew to the newsstand where he had purchased the ticket. And he bought another one. With the camera going for a close-up, he set about scratching the new ticket. The first square read $100,000. He scratched another. Again it read $100,000. You need three squares showing the same amount to win. He scratched the third, and it was another $100,000. He became dizzy with disbelief. This must surely be one of the most amazing pieces of luck ever caught on camera.

Wet, Wet, Wet

In November 1969, a newspaper reported an unusual near-drowning of eight people. The drama began with an early evening walk. John Dunne and his wife, Nora, were walking along a quay. They strayed too close to the edge and part of the cliff suddenly fell 65 feet into the sea below. Because of this, one of them must have stumbled. In an attempt to prevent an accident, the other must have reached out to stop the fall. Unfortunately, both Mr. Dunne and his wife fell into the sea. Neither of them could swim.

Luckily, Nora's brother and his girlfriend were following close behind. They both dived in after the pair. But neither of them could swim either. All four of them cried for help and attracted the attention of two police officers. The police officers immediately dived in to rescue them. This would have been helpful, except that the police officers couldn't swim either.

A large crowd gathered as all six people cried for help. Two men in the crowd, who were both strong swimmers, dived in. But the effort to rescue so many people who couldn't swim was too tiring and difficult. They were soon in trouble themselves.

By now the crowd had swelled and people were offering advice through a bullhorn. One person snatched the bullhorn and just yelled at the drowning people to stop fooling around.

The fire department arrived and threw safety lines to the people. All of them were brought safely to shore. The only two who suffered any injuries were the two men who were strong swimmers. They were taken to the hospital.

Relief at being rescued soon turned sour. The remaining six got into an argument with the crowd about what had happened and who was responsible. It's amazing what can happen on a simple evening walk.

Wouldn't They Run Out of Names?

When a woman gives birth, what are the odds of the baby being a girl? It would be logical to assume that there would be a 50-50 chance. We can't ever predict with any certainty the gender of any baby. But in some cases our chances are better than others. The Pitofsky family in New York, for example, can safely bet that the next child would be male. Unbelievably, seven generations of this family have produced only sons. In 1959, in Scarsdale, New York, the 47th consecutive son was born to a member of this family. Mathematicians have worked out that the odds of this happening are 136 trillion to 1. That's incredible!

Two of a Kind

Lots of stories of extraordinary coincidences surround identical twins. The story of the Jim twins is almost unbelievable. These identical twins were separated at birth and brought up in different cities. Their mother couldn't afford to care for them. So she gave them up for adoption. They both married girls named Linda, and both marriages failed. Each then remarried girls named Betty. When a son was born to one of the twins, he named him James Alan. The other twin named his first boy James Allen. They each had a dog named Toy. And each of them enjoyed woodworking and suffered from bad headaches. Just coincidence?

This Can't Be True, Can It?

What are the chances of meeting someone with exactly the same name as yourself? Supposedly, a man named Jim Hunkin fell into the water near a small town in England. It was reported that the name of the man who threw him a life preserver was also Jim Hunkin. Weird? A third man, also named Jim Hunkin, pulled him into a boat and saved his life. The men were not related and had not met before.

Now That's Hard to Swallow!

A sore throat is annoying and uncomfortable. Dick Winslow's throat was so sore that he went to a hospital in Los Angeles. He wanted to see if something could be done about it. He felt a lot better after going there. The doctors removed a Mickey Mouse watch that had somehow gotten stuck in his throat.

3

Introduction

Imagine: being dared to go into a spooky place alone.

"It's haunted."

"An old witch lives there."

"Someone was murdered in there."

Julia Masters listened to her friends as they walked past the old house on their way to school. Every morning they said the same things. They always crossed the street before they reached the house. But sometimes they would stop and stare at it from a distance.

If Julia was on her own, she never stopped. She hurried past the house as quickly as she could. It scared her. There was something spooky about it. The garden was overgrown, and the trees threw gloomy shadows across the porch.

Julia's father said that old Mrs. Ford lived there. He hadn't seen her in a while, so maybe the place was empty now.

"Oh, she's still there," said Mrs. Masters. "I see the visiting nurse stop by from time to time."

"Is she a witch?" asked Julia.

Her parents laughed. "Is that the latest rumor? People used to say that she was very rich and never got over the death of her husband."

"Did she kill him?" asked Julia, her eyes wide.

"Maybe," said Mr. Masters and winked at his wife.

The next day, Julia was able to add this to the other stories.

"She killed her rich husband and chopped him up." She thought this made sense after talking with her parents.

"I dare you to knock on the door," said Meg.

"I double-dare you," said Jessie.

A double-dare. Julia was trapped.

"All right," she said. "Tonight. You all have to sneak out and meet me here at 9 P.M."

They gasped but nervously agreed.

There were no clouds that night. A bright moon deepened the shadows around the old house. Julia's parents were talking with her dad's boss and thought she was upstairs reading. None of Julia's friends showed up. It was so typical of her friends. They were brave in sunlight and scared in moonlight.

"Cowards," she said to herself when it was clear they had chickened out. They'd all have excuses tomorrow.

Julia was about to return home when a band of yellow light caught her eye. A curtain in the front room had moved, revealing for a moment the light beyond. Julia, suddenly more curious than afraid, crossed the street. She hesitated for a moment at the gate and then cautiously pushed it open. It moved soundlessly on its hinges. As soon as she had crossed the spooky yard, she began to shiver, despite the warm evening. She even felt a little dizzy. Every stupid horror film she had ever seen began to crowd her imagination. The slight rustling of the leaves in the garden made her blood run cold. Taking a deep breath, she moved on.

Julia's foot had just touched the porch step when a sound made her freeze. It was a voice coming from inside the house.

"Help," it said. "Help. Somebody."

Julia panicked. She ran home and told her parents what she had heard. There was no way she was going to investigate on her own. This might have been a way to get her inside the house.

Julia's parents were furious that she had sneaked out after dark. Their anger was mixed with pride, though. If Julia hadn't been at Mrs. Ford's house, the old woman would almost certainly have died. She had fallen and broken her hip. The nurse was not due for another two days.

Julia wondered what strange twist of fate had brought her to Mrs. Ford that night. She met her in the hospital. She was an old lady with a kind face. Julia did not confess why she had come that night. . . .

3 Noble Failures

Julia Masters was in the wrong place at the right time. We've all heard about people who have been in the wrong place at the wrong time. Something unpleasant usually results from this.

However, it is also true that sometimes events that look bad may result in positive outcomes. Mrs. Ford was lucky Julia sneaked out that night. This chapter looks at some events that had unexpected results.

The Leaning Tower of Pisa

The Italian town of Pisa is a popular place for tourists to visit and take photographs. They go to look at one of the most famous mistakes in the world—the Leaning Tower of Pisa. What went wrong?

In 1173, construction began on a bell tower for the Cathedral of Pisa.

Three stories had been completed. But it soon became obvious that the ground underneath was wrong for such a heavy building. When the tower began to lean slightly, it was decided to stop building.

The tower was not touched for 100 years. In 1275, a new generation of builders made the decision to try again. It took another 25 years for three more stories to be added.

The tower continued to lean. The builders continued to build. In 1350, it was eight stories high and beautiful—but not perpendicular.

Over the next few hundred years, the tower kept leaning further. It moved only 1/25 inch a year. This doesn't sound like much. But the tower now leans at an amazing 16 feet.

Tourists Love the Error

It is because the Leaning Tower of Pisa is not right that tourists come from everywhere to see it. No one wants it to fall over. And no one wants it straightened up. Without a leaning tower, the town of Pisa would lose its biggest source of income. However, something had to be done to stop the lean from getting worse. Obviously, unless a way was found to stabilize the Leaning Tower of Pisa in its current position, it would eventually fall over.

In 1934, cement was injected under the tower. This made the tower lean even more. In 1995, someone had the bright idea of freezing the ground around the tower. The idea was that if the ground was frozen hard, the tower would stop sinking into it. The engineers who came up with this idea checked it the following morning. They found that the tower had moved another ten inches overnight. Disaster! In one night, the tower moved 250 times more than it would have in an entire year.

Heavy lead weights were attached to stop any further decline. The latest plan is to attach cables to a giant hook. Then the tower can actually be pulled back slightly. If it can be moved back toward the vertical a bit, this would slow down its lean. Engineers could probably straighten it completely. But that would ruin one of the most beautiful mistakes in history.

Christopher Columbus

Christopher Columbus may have been a brave explorer, but he wasn't a fantastic navigator. He made a mistake when he first sighted the land that forms part of the Americas. He thought he had reached India.

In fact, there are quite a few things about Christopher Columbus that need clearing up. First of all, there's his name. He wouldn't answer you if you said, "Hi, Christopher Columbus." He might have answered to Cristoforo Colombo, the Italian form of his name. You'd definitely get his attention with Cristobal Colón, his Spanish name.

That's not important, though. What is important is his mistake. It revealed to the rest of the world the existence of a land that would become North America.

You might be surprised to know what Columbus thought he was doing. He wasn't trying to prove that the earth was round. He already knew that. The idea behind his expedition was to find a fast way to China and India where Europeans could trade for silk, spices and other goods. Because he knew that the world was round, he calculated that he could do this by sailing west. He had trouble raising money for his expedition. Most people thought that it was simply too far away.

After doing some research, Columbus decided that the earth was smaller than previously believed and mostly land. Based on this, Columbus worked out that Asia was only 1,440 miles away. It would not be too expensive to outfit ships with supplies to cover that distance. Queen Isabella of Spain agreed to finance the expedition. Of course, Columbus's calculations were completely wrong.

The Journey Begins

On August 3, 1492, Columbus set out. He had only three ships with him, the *Niña*, the *Pinta*, and the *Santa Maria*. On October 12, land was reached. Columbus set foot on an island in the Bahamas. When he met the local people, he called them Indians because he thought he was in India.

Columbus made three more voyages. Up until his death, Columbus believed that the places he visited were a part of India. His ideas about reaching land by sailing west were correct. He just wasn't where he thought he was. Nevertheless, his explorations opened the way to a whole new land that would become the United States.

Accidental Inventions—Vulcanized Rubber

Rubber is a useful substance. But it easily cracks when it is cold, and it is sticky when hot. For a long time, people experimented with ways to prevent this from happening.

Inventors had unsuccessfully experimented with mixing sulphur and rubber. In 1839, Charles Goodyear accidentally dropped a piece of rubber treated with sulphur onto a hot stove. To his amazement, he found something wonderful happens when sulphur and rubber are heated together. The rubber becomes much stronger. This process is called vulcanization. It remains the basis of how rubber is made today. Unfortunately, Goodyear did not benefit from his discovery. He died in poverty.

The Stump-Jump Plow

In the early days of settlement in Australia, clearing the land to grow crops was a major problem. The land was covered with a densely growing type of eucalyptus tree. The roots of these trees are made up of several thick stems. These grow out of the stump at, or just below, ground level.

The stumps were hard to remove, so they were usually left in the fields. This made plowing difficult and dangerous. The plow could catch on the stumps and be damaged. Or the horse pulling the plow could be injured.

A man named Richard Smith was plowing his field one day. The piece of machinery he was using hit a stump, and a bolt broke. To his surprise, the blade of the plow did not stop working. So he kept on plowing. When he came to a stump, the blade just lifted up over it. Then it fell back to the ground on the other side. This was because the board to which it was attached had broken loose.

This accident gave Smith an idea. With help from his brother, Smith invented a machine that would slide over stumps when it hit them. This invention allowed the opening up of land in the southern parts of Australia. Use of this machine spread throughout the world and was particularly helpful in areas covered with small shrubs.

Anesthetics

Ether was believed to have been discovered around the 13th century. If people sniffed the gas in small doses, it made them giggle.

In 1923, a doctor in England noticed that the gas made his flowers wilt. Wondering what effect the gas would have on him, he breathed in a good lungful and passed out. After he woke up, he realized that it had great potential as an anesthetic. Soon after, ether became the most common anesthetic used in operations.

Digestion

Digestion is the process of breaking down food so it can be used by the body. In the days before x-rays and modern technology, doctors had never been able to observe digestion. A soldier who had been badly wounded provided doctors with the first chance to do so.

He had been shot in the stomach, and his injury did not close up properly. This enabled doctors to watch his intestines as he ate. Apparently he was quite comfortable and was not in pain. There was nothing between his insides and the outside world, that was all. This wound provided invaluable information for the study of the stomach.

An Explosive Discovery

Gunpowder was the only explosive known for a long time. In 1846, as the result of an accident, the first of the modern explosives was discovered.

The accident happened to a German chemist named Christian Friedrich Schonbein. He had annoyed his wife many times with his bad-smelling chemicals. Therefore, she had told him that he was not allowed to use her kitchen for his experiments anymore.

Even with his wife's warning, he still did his experiments in the kitchen when she was not at home. One day he spilled some nitric acid on the kitchen floor.

He quickly mopped it up with her apron. Then he hung the apron over the fire to dry and hoped his wife wouldn't notice.

To his surprise, there was an explosion. He realized that the nitric acid had reacted with the cellulose in the cotton apron to form nitrocellulose. This new explosive soon replaced gunpowder in weapons.

4

Introduction

Imagine: falling into a place of total darkness and being injured.

I wasn't aware of any pain. Not at first. The first thing I thought was that my parents were going to be really mad. How many times had they told me not to go too near the edge of a mine shaft?

"It's dangerous, Karen," Mom would say. I used to roll my eyes. I was sick of hearing about how dangerous it was. I knew it was dangerous, and did she think I was stupid? As if I would ever fall down a mine shaft. Famous last words.

This is how it happened. We live on the edge of a national park. This park is an area where people used to mine for gold about 100 years ago. There are shafts everywhere. You have to be really careful walking through the woods. Not all of the entry points have been discovered. There are places where these large holes are all overgrown. If you're not careful, you could easily get swallowed up. I would never ride a horse through this park, although many people do. I've lived here all of my life, so I know what the dangers are. That's why I got so annoyed when my parents kept reminding me. I was more careful than they were. They'd go off bird-watching and be looking up instead of watching where they were walking. They hated it when I reminded them of the time Dad twisted his ankle in a pothole.

I was walking in the park after school. There's this really cool mine with a rusty, tin shed around the entrance. It's a big hole, but the edges are really sharp. Like the ground just falls away suddenly into blackness. I once shined a flashlight down there, and the drop goes on forever.

I heard this scratching noise and saw a raccoon walk up to the mine entrance. I stood still and then followed as quietly as I could. I was concentrating hard on being quiet. I didn't notice how close I was to the edge. Then the earth just gave way under my weight. I remember the sickening feeling of falling. My arm hit something, and I heard an awful sound like a twig snapping. The next thing I knew, I was lying on a ledge in pitch darkness. The only sound was my breathing.

The pain in my arm grew quickly after the first shock wore off. I knew my arm was broken because I couldn't move it. Also it felt as if it was lying at a strange angle.

All I could think about was how much trouble I was in with my parents. They probably wouldn't allow me out of the house after this. I couldn't see the sky from where I was, so I had no idea of the time. How long had I been unconscious? Were people already looking for me? I didn't feel hungry, so I thought I'd probably only been there for a few minutes. On the other hand, I felt sick to my stomach. I wasn't sure whether I was hungry or not. I called out.

"Help! Help!"

My voice echoed, and I immediately felt really weak. The effort to call out used up all of my strength.

I passed out again. It was the sound of voices that brought me around.

"She's here!"

The beam of a flashlight shined down on me. Then I heard my mother's voice.

"Karen! How many times have I told you to be careful?"

It was the sweetest sound I have ever heard. . . .

4 Survival Against the Odds

Surviving a disaster takes a mixture of extraordinary courage and incredibly good luck.

In the story, Karen has a lucky escape when her fall is broken by a ledge. She was doubly lucky because she was found quickly. The stories that follow are about people who have been through experiences most of us can hardly imagine. We are astounded by their endurance, courage, and amazing ability to keep hope alive. It makes us wonder how we would react in a similar situation. They are ordinary people made extraordinary by the things that happened to them. Perhaps the lesson we can learn is that there is something extraordinary in all of us.

Tony Bullimore

Tony Bullimore was an experienced sailor. He loved the sea. There was one race that he was determined to sail in: the Vendee Globe. This race was a solo, nonstop, around-the-world yacht race. Tony Bullimore saw this race as one of the biggest challenges of his life. He could not know that he would meet with a disaster. Or that it would almost kill him.

Tony Bullimore knew so much about boats that he was able to build his own yacht for the race. His boat, the *Exide Challenger*, was designed and built with the knowledge that it had to be good. His life would depend on it. The keel of his boat was the only part not made by him. On January 5, 1997, this keel snapped off in high seas, and Tony Bullimore's life changed forever.

The Treacherous Southern Ocean

The waters of the Southern Ocean are freezing, and they are often violent and unpredictable. It takes courage to sail them alone in a yacht. At 7:31 P.M., Tony was below deck in the galley. He was dry and warm, but the seas were wild. Imagine what it must feel like to be trapped inside what is little more than a tiny room—and on top of that being tossed around like a rubber ball.

Suddenly, the *Exide Challenger* was lifted by a huge wave. It then began a rapid descent down the face of the wave. As it plummeted, Tony heard a loud noise. He realized that his yacht had hit something at high speed. In the next few seconds, the boat overturned and everything went black.

The ocean threw the upside-down yacht about like a cork. Inside, Tony Bullimore was trapped in a tiny space between the ship's wall and the ocean outside. This was the only place where there was any air. Beneath him, in the cabin, the sea sucked in and out with the force of a waterfall. He could see nothing, and the temperature was close to 32°F (0°C). The noise was absolutely terrible. How was he going to survive?

He thought that his best chance was to somehow get to the life raft. Then he could float away from the capsized yacht. At least an aircraft would then know that he was alive. But the problem was that he had to swim in pitch darkness. The life raft was attached to the deck that was underwater. He tried this 12 times, and each time he failed. Each time he had to swim back to his air pocket and try to warm up. If he hadn't been wearing good quality clothing, he would have died of the extreme cold.

Apart from hunger and thirst, perhaps his greatest enemy was the cold. If he began to suffer hypothermia, he was lost.

He had to make a decision. Should he stay inside his upturned yacht and risk rescuers thinking that he had perished? Or should he go outside and cling to the boat where he might be seen? He had frostbitten toes because he was unable to keep his feet out of the freezing water. Also, he had lost a finger when a hatch slammed shut on his hand. He was hungry, thirsty, and very, very cold.

He decided that his best chance was to stay inside. As it turned out, this was the right decision.

At 11:42 P.M. on Wednesday, three days after the disaster, Tony heard an electronic ping. A search plane had seen his yacht and dropped an electronic probe. Tony began tapping on the boat. The probe would transmit his tapping to the rescuers. Then they would know that there was someone alive inside the upturned boat.

Saved

At 12:22 P.M. on Thursday, the naval ship *Adelaide* located the yacht. It sent an inflatable rubber boat to see if Tony was still alive. He heard the rescuers knocking on the outside of his boat. He called to them, but they couldn't hear him. He decided to dive out from under the yacht. Taking a deep breath, he dived and swam to the surface.

When he emerged, the first thing Tony saw was the *Adelaide*. "When I looked over at the *Adelaide*," he said, ". . . I was looking at life. I think if I was picking words to describe it, it would be a miracle. An absolute miracle."

Tony Bullimore's story is remarkable because it reveals how important it is to never give up hope.

The Remarkable Icewoman

In January 2000, it was reported that a Norwegian woman had been brought back to life. This happened after she was clinically dead for more than two hours.

She had been skiing with friends when she fell and became wedged between rocks under over-hanging ice. She was drenched again and again by icy water. The situation was hopeless. She couldn't move, and her body temperature began to fall. Her heart and breathing stopped before rescuers arrived.

Never Say Die

There were no signs of life when she was taken to the hospital. Her body temperature had dropped to 56.7°F (13.7°C). No one had ever survived such extreme hypothermia. Indeed, if our body temperature drops below 82.4°F (28°C), our chances of survival are really slim.

When the woman arrived at the hospital, doctors tried something extreme. They removed all of her blood. Then they warmed it and put it back. Incredibly, the woman came back to life. There were concerns that she might have suffered brain damage. But her brain had cooled to the point where the lack of oxygen did not do it any harm.

Her struggle wasn't over though. At first she was paralyzed. And she spent many weeks on a machine that did her breathing for her. But eventually, the woman recovered completely. She couldn't remember the accident, and she started skiing again.

James Scott, the Iceman

Have you ever gone for a whole day without eating anything? Imagine going without food for 43 days!

One of the most remarkable stories of survival is about James Scott, a 22-year-old medical student. He survived in the mountains of Nepal for 43 days.

He had no food and no water. How did he manage to stay alive?

In his final year of studying medicine, James Scott thought he needed a break. He was a fit young man, so he decided to go on a major hike. James and his friend Tim headed off to Nepal.

A Bad Mistake

On their first night, James and Tim met some people at a lodge. After talking with the other hikers, they decided that the hike they planned on was too easy. They were given a map of a more challenging trail. James and Tim met another man, Mark, who agreed to join them.

Not long after starting, Tim's knees became painful. He decided to turn back. James decided to go on. He made two serious mistakes at this point that almost cost him his life. He gave Tim the map, and he did not ask for the lighter that was in Tim's bag.

The first night on the new trail was cold. The following morning, light snow covered the roof of the lodge they had stayed in. James and Mark were concerned. But they were told that no more snow was expected that day. This turned out to be wrong.

Another Bad Mistake

After walking for several hours, James and Mark began to feel nervous. The temperature dropped so low that the camera jammed. Then the snow began to fall heavily, and within minutes the trail disappeared. They discussed what to do. James knew they were in trouble. Mark decided to push on to reach the end of the trail and go down the other side. James decided to return the way they had come. They separated—another mistake.

James thought that he recognized where he was. But snow makes everything look the same. He became hopelessly lost. He wasn't wearing the right clothes for the weather conditions. The only food he had with him was two chocolate bars. He was wet, cold, and exhausted. By now it was getting late. He found a place to spend the night. He toweled himself dry, knowing how dangerous it was to be wet in such conditions. He changed into dry clothes and curled up in his sleeping bag. He also ate some of his chocolate, which was his only meal since breakfast.

Lost

The following morning, he set out. But the more he walked, the more lost he became. That night, he found a large, rocky overhang where he took shelter. He dried himself as well as he could and tried to sleep. He felt so cold that he thought he would probably die of exposure that night. He wrote what he thought was a last letter to his family, telling them he loved them. He then ate the last of the chocolate.

He woke and his third day began. His feet were numb. When he took his socks off to check them, his feet were blue. He was thirsty and ate snow. He soon began shivering uncontrollably and realized that the snow had lowered his body temperature. He passed out. Then the struggle of waiting to be rescued began.

How Could Anyone Survive This?

Some days were sunny, and there was water where snow melted. The hunger was terrible. As a doctor, he knew what was happening to his body. It was feeding off the protein in his muscles. He was wasting away. Then one day, he wasn't hungry anymore. This was not a good sign. His mouth was full of ulcers. He could not feel his legs, and he began vomiting. The hunger returned. Snow fell heavily. He thought of taking his own life and then thought of his family. He couldn't do it.

On the 43rd day, James heard a helicopter. His sister, who never gave up hope of finding him, had organized one final sweep of the mountain. He stumbled into the open and waved. He couldn't see anything because, by this stage, his vision had been affected by starvation. Even if they had seen him, it would take them a long time to reach him. James felt that he was within hours of dying. He returned to his sleeping bag.

Unbelievable

Many hours later, two Nepalese guides found James. They had risked their lives to rescue him. They told him that no one had ever lived more than ten days in these mountains. Later, the full search party arrived. The following morning a helicopter lifted James to safety. As he was carried up to the helicopter in a harness, James looked back at where he had been. He still thought that this was one of the most beautiful places on earth.

Caves Are Dangerous Places

People love to explore caves. Caving is a sport that grows more popular each year. But it is very easy to get lost in a cave. This is true even if you are experienced and with a group of people.

In Utah a man named Josh was exploring a cave.

He was with a group of people that included his father. Josh was not carrying a flashlight, which was not a very good idea.

Somehow he lost sight of the group ahead of him. One moment he could see them up ahead. The next, he was in pitch darkness. They didn't hear his calls. Josh decided to go back the way he came by feeling his way along the wall. He assumed he was going in the right direction. But he must have taken a wrong turn at some point along the way and got lost.

What is really incredible about this story is that Josh ended up being lost for five days. This is despite the fact that he had been with a group. And the group went back to find him right away when they realized he was missing. Imagine living in total darkness without food for five whole days.

The Trip of a Lifetime

It was a dream of Dougal Robertson's to sail around the world. In 1972, his wife, Lyn, and their three sons, Douglas, age 18, and Neil and Sandy, twins age 10, set out. A young man named Robin Williams was also on board to help the family. On the night of June 15,

1972, their dream became a nightmare when their boat, *The Lucette*, sank. All on board managed to make it into the rubber raft safely.

The raft had a survival kit. It contained bread and water that might last a few days at the most. There were also flares, fishhooks, bellows for keeping the raft inflated, and oars. On this small raft, they floated at the mercy of the sea and winds somewhere in the Pacific Ocean.

One of the hardest things about being on the ocean is that there is water all around. But this water is undrinkable. So water quickly became the most urgent need for the Robertson family.

Water, Water

Whenever it rained, a rubber sheet was used to collect and funnel water into bottles. This made the water taste like rubber. It did not rain often. To keep from dying of thirst, the group was forced to drink the blood of turtles. Turtles also provided meat for the family. The only other food was the occasional flying fish that landed accidentally on the raft. Sometimes they happened to catch a fish as well.

The Robertsons were rescued on the 38th day of their ordeal. Their story is an incredible tale of survival against the odds. They had all lost weight and were suffering from dehydration and lack of sleep. But apart from this, they were in very good condition.

Introduction

Imagine: being trapped in a small cave through a blizzard.

I never used to be a snow person. I liked the beach. Heat. Sand. Sweat. But last year I discovered a feeling for snow. I stole that phrase from a book I read. "A feeling for snow." We—that's my brother Tom and I—were staying with our cousins in a lodge. This wasn't the first time I'd seen snow. It was the first time I took skiing lessons and used a snowboard.

I never realized how hot you get when you ski. You start out freezing and end up sweating. There's nothing like the incredible rush of flying downhill. I learned to love the fear. I also learned what it meant to be really afraid—because I almost died.

The worst day of my life began on a morning that was clear and crisp. The snow glistened under an electric blue sky. It was the kind of morning that made you feel as if nothing could possibly go wrong. My brother and Jake, my older cousin, decided to take our snowboards and get away from the resort. Jake knew a slope where no one ever went. It meant going off the trails, but the day was clear and beautiful.

We put chocolates and water in our packs. We didn't take lunch because we intended to be back before noon. Jake carried a small shovel across the top of his pack. He always carried this. He was sensible and took no risks. If he wasn't responsible, our parents would never have let us go with him. Everyone knew where we were headed, so what could possibly go wrong? Plenty.

The weather on a mountain can change without warning. The wind increases and before you know it, it's tugging at your clothes. And it's blowing snow in your face. A blue sky can vanish behind dark clouds in a few minutes. I knew something was up when Jake said that he thought we should head back. We'd only just arrived, so I knew he wasn't kidding.

The weather closed in with incredible speed and strength. I couldn't tell the difference between the snow and the sky.

"It's a white-out," said Jake. "We can't risk getting lost in this. We'll build a snow cave and stay put."

Jake knew what he was doing. He dug a snow cave, and we huddled inside. There wasn't much room, and it was freezing. But we were out of the wind. We thought we'd be there for a few hours, but the weather turned into a blizzard. I was terrified that the cave would collapse under the weight of the snow building up around it. Jake said that wouldn't happen. We kept a hole near the entrance clear of snow so that air could get in. We spent the whole night there. We took turns staying awake and clearing the hole.

We walked out the next day, hungry but okay. I saw my father cry for the first time when we arrived back safely. It was an experience, all right, but I never want to repeat it. . . .

5
Life or Death?

The story of the boys and the snow cave has a good reminder. We should never take our safety for granted.

For them, survival depended on Jake's knowledge and experience. They spent an uncomfortable night with nothing to eat but chocolate. But they got through it without injury. Not everyone is so lucky. Some people find themselves in situations that push them to their absolute limits. The stories of how they survived are truly amazing.

Douglas Mawson

Antarctica is the most difficult place to live on Earth. This continent is made up of 98 percent ice and snow. Until recently, no humans lived there, and even now, no one lives there permanently. As the last of the unexplored wildernesses, Antarctica attracts many explorers.

One of the greatest of the Antarctic explorers was Sir Douglas Mawson. The story of his incredible journey across the frozen wastes is an unbelievable tale of hardship and courage.

A First

Mawson was an Australian scientist. He was not really interested in being the first or the most famous Antarctic explorer. He wanted to increase our knowledge of this large continent.

At the time his first expedition started in 1907, people had little idea of what Antarctica looked like. They hadn't seen pictures of it or read about it. Mawson joined the great polar explorer Ernest Shackleton on this expedition.

Shackleton, Mawson, and two other Australian scientists set out to find the south magnetic pole. They walked an incredible 1,200 miles, pulling sleds loaded with supplies over difficult terrain. It took them four months. But they were the first people to reach the south magnetic pole.

Having achieved this, they had to walk back. They made it just in time. If they had been late, their ship would have left without them. The ship had to avoid being trapped all winter in the ice.

In 1911, Mawson organized an expedition of his own. He was a patriotic man. He believed that it was important for Australia to claim some of the Antarctic territory as its own. The *Aurora* sailed on December 2, 1911. When it reached the Antarctic, a camp was set up at a place called Cape Denison. The conditions were unbelievably harsh. The wind was ferocious and constant. Only short explorations of the area could be carried out until Antarctica's winter ended.

The Team Sets Out

In November 1912, Mawson set out on a major scientific exploration. He was accompanied by Lieutenant Ninnis and Doctor Xavier Mertz. They brought with them a team of dogs and supplies for 12 weeks. The weather was bearable. But the journey was difficult and exhausting.

Then disaster struck the explorers. Antarctica has several hidden dangers. The biggest of these is a crevasse that can open up instantly under your feet with little warning. Lieutenant Ninnis and most of the dogs disappeared down one such crevasse. One moment they were there, and the next they were gone. To make matters worse, most of the food supply was on the sled that was lost.

Mawson and Mertz had at least 30 to 35 days worth of traveling ahead of them before they would reach the base camp. The two men had only ten days worth of supplies left. Faced with starvation, they killed the remaining sled dogs one by one for food. They ate the livers, believing them to be the best part to eat. Dog livers, however, contain a very large amount of vitamin A, which is poisonous in large quantities. Mertz soon became very ill. He was so delirious that he chewed off one of his own infected fingers.

They pushed on, with Mawson, who was himself very ill, pulling Mertz on a sled.

What Now?

On January 8, Mertz died. Mawson buried him in the ice. Then he had no choice but to wait three days for a period of darkness to pass before he could struggle on. At one point, he fell down a crevasse. He was saved because his sled caught on a piece of ice. Using valuable energy, he climbed to safety. He could only walk a couple of agonizing miles each day in the unbearable cold. It was only his strong will that kept him going.

On January 29, 1913, he found food buried by a search party. He realized that someone was out looking for him. He did not know that he had missed them by only a few hours.

Mawson knew that he was approaching his final destination. The main camp was close now. The food restored his energy and his determination to survive. But the weather suddenly turned very bad, and he had to wait again. He knew that the ship was waiting for him. But it could not wait for long, or it would run the risk of becoming iced in. If he missed it, he faced another winter, this time alone, in Antarctica.

In the End

He thought his luck had run out. He reached the camp, only to see the ship disappearing over the horizon. It had left a few hours earlier. He was not alone, though. Six men had remained behind, waiting for him. They waited out the winter with Mawson.

Mawson's refusal to surrender to death is inspirational. It is difficult to imagine what it feels like to be truly cold, exhausted, and hungry. Could you force yourself to drag a heavy sled over the ice when you hadn't eaten for days?

Dr. Jerri Nielsen

Antarctica in winter is one of the most hostile environments on Earth. Several countries around the world, however, maintain bases in Antarctica. During the long winter, the people who work there are isolated from the rest of the world.

In the winter of 1999, at an American base, Dr. Jerri Nielsen detected a lump in her breast. It was impossible for her to leave the base. She faced the possibility of having to deal with breast cancer without proper medical assistance.

An Amazing Mission

The most important thing to do was to establish whether Dr. Nielsen would need chemotherapy. It was decided to drop the necessary supplies and equipment to the base. This involved a high-risk flying mission. The temperatures were far below freezing. And there was complete darkness (the sun doesn't rise during the Antarctic winter). It was considered too dangerous for the aircraft to land. So the supplies had to be dropped. A U.S. Air Force aircraft set out from New Zealand. This was to be a 6,696-mile round trip. The plane had to refuel in mid-air during the night. When the aircraft reached its destination, the winds were blowing at 180 miles per hour. The temperature was -94°F (-70°C). This was the first mid-winter drop ever made over Antarctica. It was successful. But for Dr. Nielsen, the ordeal had just begun.

Among the equipment was a digital microscope. Nielsen had to take a needle biopsy of her own breast. Then she had to examine the tissue sample under the microscope. With communication technology, the results could be analyzed by experts in the United States. The analysis was positive: the tissue was cancerous.

Following the advice of specialists via satellite, Dr. Nielsen began administering her own treatment. This took enormous courage and determination.

Imagine what it must have been like for Dr. Nielsen. She had to treat herself with medications that made her feel very ill. She needed to get out of Antarctica for proper medical attention but she was trapped. This was what Dr. Nielsen bravely faced that winter.

An Even More Amazing Mission

Dr. Nielsen endured the long winter, but it was vital that she get out as soon as possible. In October 1999, a U.S. military crew performed one of the most daring rescues in the history of Antarctica. Spring had just begun, but the weather continued to be extreme.

Using a ski-equipped Hercules aircraft, it was decided to attempt a landing at the base. This would be the earliest that such a landing had ever been made.

In extreme conditions such as these, landing gear and wing flaps can fail. The lowest temperature for the Hercules to safely land is -58°F (-50°C). On the day of the landing in 1999, the temperature was even colder, -59.8°F (-51°C).

The mission was successful, and Dr. Nielsen was taken for proper medical care. Her mother said that her daughter liked adventure. She said that courage was one of her daughter's finest qualities. It had taken all of her courage to face cancer in one of the most remote places on the earth.

The World's Unluckiest Woman

Some people just seem to be accident prone. You probably know someone who always has a bandage covering the latest injury. The all-time record for surviving accidents must belong to a woman named Mary Regere of Santa Rosa, California. In the 1930s, it was reported in *Ripley's Believe It or Not* that Mary survived 13 car crashes. The cars were totally destroyed. You'd think she'd stay away from cars after the first few accidents, wouldn't you? She was called Calamity Mary because of her history.

She didn't have much luck with the animal world, either. She was thrown from a horse, bitten by a dog, kicked by a cow, and mauled by a cat.

Even the weather was against her. She survived a hurricane, a cyclone, an earthquake, and a flood.

Incredibly, she thought it might be safe for her to fly somewhere. The plane she was flying in lost power and dropped 10,764 feet before righting itself. Mary was uninjured.

Indeed, she suffered very few injuries in her life. Still, if Mary called you and asked you to go for a drive, what would you say?

Apollo 13

Scientists are not, as a rule, superstitious people. It would have seemed silly to name the 13th Apollo flight *Apollo 12.5.* Or to skip the number altogether and call it *Apollo 14.* After all, 13 comes after 12, and that's that. But even the most doubting scientists must have wondered about unlucky 13. Especially after the disasters that happened to flight *Apollo 13.*

The first landing mission to the moon, *Apollo 11,* took place on July 20, 1969. *Apollo 13* was launched on April 11, 1970. It had a crew of three and was to be the third moon landing mission.

Apart from the rockets, the Apollo spacecraft is really two separate spacecrafts. They consist of the Command Service Module (CSM) and the Lunar

Module. The CSM is the craft that remains orbiting the moon. The smaller Lunar Module lands on its surface. At the end of the lunar exploration, the Lunar Module returns to the Command Service Module. This module then re-enters the earth's atmosphere and splashes down safely. Unfortunately, things didn't go quite that way for *Apollo 13*.

Houston, We Have a Problem

Just before the crew began preparations for the lunar landing, they heard an explosion. It was in the Command Service Module. You can imagine how terrifying this would be. They were crammed into this tiny area, hurtling through space with no hope of rescue. Suddenly they were without oxygen or power. There was only one thing to do. They had to squeeze into the Lunar Module.

The Lunar Module was not equipped for a lengthy period in space. There wasn't enough oxygen or water to last more than a couple of days. The astronauts' only hope was to change the course of the Command Service Module. They needed to head it back toward Earth. But they couldn't just detach the Lunar Module and fly it back to Earth. It didn't have a heat shield. The Lunar Module would burn up as soon as it hit Earth's outer atmosphere.

All of the systems that were still working in the CSM were shut down. This was done so it would have enough power to get back to Earth. Imagine how desperate the families of the astronauts must have been. Every time they looked up at the moon, they must have been wondering and worrying about their loved ones.

A Tricky Maneuver

Inside *Apollo 13* it was decided that a tricky maneuver would be attempted. First, the landing engine of the Lunar Module would be fired. This would provide enough force to push the Command Service Module out of its lunar orbit and then propel it toward Earth.

All of this was dangerous and complicated. Temperatures inside the Lunar Module were freezing. The air the astronauts were breathing had become dangerously high in carbon dioxide. In high concentrations, this is a poisonous gas.

While the world watched and waited, *Apollo 13* approached Earth's atmosphere. On April 17, 1970, *Apollo 13* splashed down in the Pacific Ocean. The three men on board, James Lovell, John Swigert, and Fred Haise, were safe and well. The adventure of *Apollo 13* was later made into a film of the same name in 1995.

An Old Superstition

Myths about the number 13 have been around for a very long time. One of the first recorded stories is an old Norse legend. There was a great feast at which 12 of the most powerful Norse gods were guests. One of the gods that everyone loved the most, Balder, was there. At some point an evil spirit called Loki entered and so raised the number of guests to 13. Loki created trouble, and Balder was killed. Loki was eventually punished for the murder.

Whatever the origin of the superstition or fear of the number 13, it refuses to go away. When someone is holding a dinner party, it is thought best to avoid having 13 guests. The first person to rise from a table with 13 guests will die within a year. That's a superstition strong enough to spook any party host.

In some cities, the number 13 doesn't appear on houses. Some hotels don't have a 13th floor or a room 13 either. They use 12a or go right to 14 instead. A famous hotel in London places a carved black cat on a seat where only 13 people are sitting. In England the black cat is thought to bring good luck.

6

Introduction

Imagine: finding a coffin floating in the cold ocean waters.

You might have heard this story before. Some people believe it's true, but it's a bit too strange for me. There was this actor in the late 19th century. He was quite a famous actor named Charles Coghlan. He was a stage actor, of course. There were no movies or television back then.

The story really begins in Galveston, Texas, in 1899. Like everyone else, Charles Coghlan was looking forward to seeing a new century begin. Luck was against him. He fell ill and died—that was the end of his New Year celebrations.

He was buried in Galveston in a lead coffin. The coffin was put in a vault. That's like a big tomb. You had to be rich to be buried in style like that. A year later, Galveston was hit by a huge hurricane. On September 9, 1900, Galveston was blown apart. The city was devastated. More than 700 people died. No one had ever seen a hurricane as bad as this one.

The cemetery where Coghlan was buried was flooded. Water rushed through and destroyed everything in its path. Lots of coffins were washed out of the graves. Most of them were recovered and reburied. But the story is that Coghlan's coffin floated out to sea. It began an incredible and unbelievable journey.

This is where things get really weird. The coffin, because it was lead, wasn't damaged by the water. It got caught in the ocean currents and was tossed about, but it didn't break open. The remains of Charles were safe and dry inside.

Galveston is near the Gulf of Mexico. Beyond the Gulf of Mexico is the Atlantic Ocean. Coghlan's coffin floated out of the Gulf and into the Atlantic. From there it floated north, all the way to Canada. That's a long way.

"Yeah, so what?" you say.

Well, here's the thing. Eight years after the hurricane, some men were fishing off Prince Edward Island, a small island off Canada. They found the coffin and brought it ashore.

Coghlan's name was on the coffin, so there was no trouble identifying him.

"Yeah, so what?" you say again.

Incredibly, the people who lived on Prince Edward Island knew Coghlan. He actually had a vacation house there. That's strange enough. But what's really strange is that Coghlan was baptized as a baby in the only church there. It was as if his coffin had made a long journey home. He was reburied in the churchyard where he had been baptized. . . .

6 Can That Be Right?

The story of Charles Coghlan is not pure fiction. There really was an actor named Charles Coghlan.

This story about the amazing journey made by his coffin was told over and over as fact. Some people were so astonished by it that they began to examine it more closely. Their research came up with some interesting facts. There is no grave on Prince Edward Island bearing Coghlan's name. It is certainly true that he had a summer house there. But the rest of the story seems to be myth. The story caught on because it was an example of a coincidence that defied belief.

What the Scientists Say

Scientists say that we shouldn't be too astonished by coincidences. Since coincidences are interesting, we tend to pay attention and remember them more than other events. However, just because we are more aware of them, it doesn't make coincidences any more amazing.

Mathematicians analyze coincidences by using probability. This is a branch of mathematics that examines chance. A mathematician might say that we experience many improbable events every day of our lives. We only notice them when they are paired with a similar improbable event. Given the number of improbable things happening to us, some are bound to match.

Try the Birthday Matchup

Ask 23 people in a room when their birthday occurs. There is a 50 percent chance that at least 2 of them have the same birthday. Of course, there is also a 50 percent chance that they won't.

What if we were among the 23 people in this sample? No doubt we would be surprised if we shared the same birthday with someone else. However, because there is a 50 percent chance of it happening, it really isn't that surprising.

What if we discovered that no one in the room shared our birthday? We would not be surprised. This is what scientists mean when they say that people attach more significance, or value, than they should to coincidences. The odds are the same for a birthday matchup as they are for no matchup. Our surprise in each case should be equal. It isn't because finding out that there is no matchup is somehow not that interesting. It seems better to find out that there is a matchup.

Are there any birthday matchups in your classroom? Ask all of your classmates and find out.

The Fourth of July

The Fourth of July is the anniversary of the signing of the Declaration of Independence. In what looks like an extraordinary coincidence, three presidents died on this date.

John Adams and Thomas Jefferson both died on the Fourth of July in 1826. They had both signed the Declaration of Independence. Their deaths occurred on the 50th anniversary of the signing that was done in 1776. James Monroe died on the Fourth of July in 1831. There was also a president who was born on the Fourth of July. He was Calvin Coolidge.

Does this mean that there is something mysterious about presidents and this date? Mathematicians would say no. It's certainly an unusual coincidence. But it's not so unusual that it can't be explained.

Heads or Tails? Try This

Here's an experiment that you can try. It illustrates how it is possible to see coincidences among random pieces of information.

Take a coin and toss it over and over again. Each time it comes up heads, jot down H on a line. Each time it comes up tails, jot down T next to the previous letter. Do this 50 times.

You will now have a line of letters. Underline all the runs with many Hs and all the runs with many Ts. Underline also any run of HTHTHT. What do you notice? Every time you toss the coin you have an equal chance of getting a head or a tail. If you have a run of heads, it doesn't mean that you must get a tail soon. You probably will, but you have an equal chance of tossing a head.

What is the point of this? If you have a run of HTHTHT, you could say, "Wow! I tossed a coin, and it changed between heads and tails six times in a row. That's incredible!" Put like that, it does sound incredible. However, it was just chance that caused the run and nothing else.

John F. Kennedy and Lincoln Again

The coin experiment shows us one important thing. It is easy to make coincidences happen by choosing which information to look at. This can be extended to include stories about coincidences in people's lives.

Let's go back to the links between John F. Kennedy and Abraham Lincoln in Chapter 1. They certainly appear to be extraordinary. But some people do not believe that there are mysterious forces at work here. They argue that there are just as many facts about these two men that are not alike. Kennedy and Lincoln weren't the same age when they died. They weren't born on the same date, or even in the same month. They weren't even born in the same state. There are hundreds of other differences between them. This shows that if we take a large enough sample of information, we will be able to see some similarities. To a mathematician, this is a simple rule of probability, or chance.

The Coincidence File

In a book called *The Coincidence File,* the author, Ken Anderson, retells a story that appeared in a newspaper. A man named Ian Heathcote regularly traveled home on a particular train. One afternoon, he felt for his wallet and discovered that he had lost it. He reported the loss to the police. He did not expect it to turn up.

Two days later, a man traveling on this same train noticed a wallet under his seat. He picked it up and looked inside. There was no money in it, only a driver's license. He looked at the photo on the license. He thought he recognized the face. It looked a bit like a man he knew at work. So he checked the name on the license.

The last name was the same as the man's with whom he worked. Incredibly, when the wallet was shown to his co-worker he said, "Hey, that's my brother."

Ian Heathcote lost his wallet and had it returned to him by his brother. What are the chances of that happening?

Glossary

anesthetic a substance, such as a gas, that causes loss of feeling or awareness

bellows tools used to produce a burst of air

biopsy the removal and analysis of cells to check for cancer

capsized turned over in the water

cellulose the main part of the plant cell wall. It is used to make products, including cotton and rayon

chemotherapy the use of chemicals to treat cancer

consecutive all following in order, one after another

crevasse a deep crack in a glacier or other surface

dehydration loss of water or body fluids

delirious mentally confused and disturbed

exposure a state or condition of being unprotected, especially in severe weather

hypothermia a dangerously low body temperature

improbable unlikely to happen or occur

keel a board or plate that runs along the center of the bottom of a boat. It goes from front to back and often projects outward.

maneuver a planned movement needing great skill

mauled injured or hurt badly

pharaoh a ruler of ancient Egypt

plummeted fell or plunged

quay (key) a structure built along water where ships load and unload their cargo

random showing no clear connection or pattern

singed slightly burned and damaged

succeeded came after another in office

superstitious fearing something because it seems unlucky or is unknown

synchronicity events that happen at the same time and seem related

terrain the visible characteristics or look of an area of land

tyrant a cruel ruler

ulcers open sores that do not heal easily

vulcanization the process of treating rubber or plastic with chemicals in order to make them more useful

Index